THE NATURE OF MASS

THE NATURE OF MASS

The Technical Version

MARC E. KING

Author of Cold Fusion

Copyright © 2013 by Marc E. King.

ISBN: Softcover   978-1-4797-8346-5
         Ebook   978-1-4797-8347-2

All rights reserved. No part of this book may be reproduced or transmitted in any form or by any means, electronic or mechanical, including photocopying, recording, or by any information storage and retrieval system, without permission in writing from the copyright owner.

This book was printed In the United States of America.

To order additional copies of this book, contact:
Xlibris Corporation
1-888-795-4274
www.Xlibris.com
Orders@Xlibris.com
129315

Table of Contents

Foreword by the Author 7
Introduction 9
III. The Nature of Mass 11
IV. The Nature of Higher Dimensional Mass 13
V. Mass x Volume 15
VI. Mass x Boundary Volume for Hydrogen 16
VII. The Nature of Quantum Spatial States
    (Bound Energy States) 18
VIII. Intersections 20
IX. Boundary Energies 22
X. Closed Spatial Travel 23
XI. Closed-Travel Energy 25
XII. n_{MAX} (Maximum n) 27
XIII. r_{MAX}, r_C, and E_C (Closure Energy) 28
XIV. Inverse and Reverse 29
XV. Exercise 30
XVI. Tables 36
XVII. Beginning to End of 3-Dimensional Intersections 38
XVIII. Diagrams 42
XIX. Technical Conclusions 51
XX. Reference 53
    Main Manuscript (without Appendices) 53
    Technical Summary of Changing Your Mind 60
    Technical Summary of Fifth Dimension 62
    Technical Summary of Cold Fusion 63
    Appendix B (Reference) 65
    Appendix C (Reference) 66
    Appendix G (Reference) 68
    Appendix R (Reference) 70
    Appendix Y (Reference) 73
    Appendix Z (Reference) 75
    Reference Texts 79

Foreword by the Author

This text assumes only space and the spatial progression shown in reference. We do not require empirical results. From dimensional space alone, this text precisely derives:

Electron Mass
Proton Mass
Neutron Mass
Elementary Mass (Allowed Atomic and Nuclear Mass)

The "charge" force and the nuclear "strong" force have been dispelled. These forces and concepts are not obsolete. They remain valuable. They represent the effective assembly-language that interfaces between observation and the machine language of 1's and 0's.

In the real world, the attributes "space and mass" in fact represent a type of "machine language" (ones and zeros) upon which everything else is compiled and constructed.

This is a technical text. Some technical readers may be uncomfortable with the technical results from this text and the preceding texts.

There is no need for discomfort.

The concept ($t=cB$) embraces all of existing physics and all science. There is no departure from our prior educations and experience.

In the same way that quantum mechanics has dealt with time and space in the non-macroscopic world, this new concept supports that same endeavor.

If you understand this material, you should be able to correctly answer Exercise 3 by yourself.

Introduction

Mass is perceived as an empirical attribute (no mass-or-mass) but having a real value, i.e. the value m (kg) defined by the magnitude of Earth gravity pulling on a mass having value m.

This ancient view is shortsighted in the modern world.

Mass is in fact defined by space, i.e. the sequence of spatial progression.

This fact is irrevocably proved, both physically and mathematically, immediately following this introduction. For the technical reader, it is best to have reviewed the texts *Changing Your Mind*, *Fifth Dimension*, and *Cold Fusion*, but necessary text from prior works has been attached to the present text as reference material.

This text will start with the straight-forward proofs required to introduce modern physics and to distinguish modern physics from the old thinking we continue to embrace.

Time t is not physically real. $t = cB$. One second of time t is a perceived representation of c events B.

To begin, we define the nature of mass without using empirical results. Mass is derived from the spatial sequence alone.

III. The Nature of Mass

Learning from the manuscript "A Mathematical Transformation of Variables . . ." (attached,) we now better understand the nature of mass.

Mass is the vehicle by which space is pulled through the natural spatial sequence.

The universal mass force is represented by the constant K(D) per Appendix W. A summary of the text *Cold Fusion* is attached as a reference.

We have seen throughout the Manuscript that the value of mass is not a function of spatial dimension. As a two-dimensional example, see Appendix B (attached.)

We also know that mass is two-dimensional in 8-dimensional space, one-dimensional in 5-dimensional space, and zero-dimensional (a point mass) in three-dimensional space. As an example, see Appendix Q (*Fifth Dimension, The Light to See.*)

We have calculated the hydrogen (H) nuclear (proton) radius in Appendix W from the zero-neutron unstable He structure:

$r_{NH} = r_{NHe} / 4 = 7.7E\text{-}15 \; meter.$

There are no "points" in real three-dimensional space:

$r_E = (3 / 4\pi)^{1/3} \times b_3 = 6.89E\text{-}18 \; meter.$

The *ratio* of "three-dimensional proton radius" divided by "three-dimensional electron radius" represents the mass force K(D) "pulling" 3-dimensional mass through 5-dimensional space per the natural sequence. Calculating exactly as in Appendix Z (*Cold Fusion*, attached) we now have the mass requirement:

$$m_P / m_E = (r_{P3} / r_{E3}) / \varphi = r_{NH}(1 + .08^{5/3}) / r_E / \varphi$$

where $\varphi = .618 \ldots$ or,

$$\boldsymbol{m_P / m_E = 1{,}836.}$$

Note:

The arguable empirical value for the ratio is 1,833 and is within the factor 0.16% (0.0016) of the calculated value. The interchange of AMU and m_P "conventional" mass values in some calculations is a likely cause for small error. As in Appendix C, we do not intend to calculate the 3rd decimal for the value 0.08 in the present context.

IV. The Nature of Higher Dimensional Mass

(Also See the Section *Intersections*)

Similarly, for 5 and 8 dimensional mass (protons and neutrons,) the required curvatures of space and the universal mass force represented by K(D) lead to the following derivation:

As in Appendix Y (*Neutron Mass* attached,) the atomic nucleus is a three dimensional intersection between five and eight dimensional space. All nuclear mass is 3, 5, and 8 dimensional at the same "time."

All nuclear mass should "tip" a three-dimensional scale.

Unlike the electron and proton masses, the proton and neutron act on the same volume of 3-dimensional space, i.e. they have the same radius $r = r_N$; therefore, $m_N / m_P = 1$.

The 8-dimensional surface (neutron) mass pulling on the 5-dimensional linear (proton) mass also must pull on the entire 3-dimensional intersection mass. Atomically,

$$m_N / (m_N + m_P) = n_N / (n_N + n_P),$$

and this defines the curvature C from 5-to-8 and 3-to-8 dimensional space for the nuclear intersection, i.e.

$$(\varphi + \gamma) / 2 \leq C < \varphi.$$

(Diagram "INTERSECTION" should help to visualize.)

Then,

$$0.500 \leq n_N / (n_N + n_P) < 0.618$$

and this agrees with Appendix Q (*Fifth Dimension*) and Appendix R (*Periodic Chart of Elements*, attached.)

Note: Standard H (hydrogen having no neutron) uniquely has curvature 0 per the attached Appendix Z. This is a special case.

Per Appendix Y, some curvatures within the above range are not allowed, and some energy states are not "100% present" exactly similar to electron "spin" energy states. The term "unstable" is inaccurate with respect to "time" but is accurate with respect to energy states.

See Table 2.

V. Mass x Volume

Mass x Volume is a useful metric and has the (J) units kg-met^3.

E / E_B is the key metric for all of quantum mechanics and has the units:

u*nit mass-boundary volume.*

The mass and volume of anything define its metric E / E_B.

Mass x Volume requires a unit change in order to convert the value to its metric E / E_B.

Considering a perfect (or nearly perfect) spherical 3-dimensional volume, such as an atomic nucleus, a planet, a star, or a black hole, then its unit-mass x boundary-volume should be represented by:

$E / E_B = 4\pi r^2 b_3$ / mass.

Except, for bound energy states ($E / E_B < 1$), b_3 may be unavailable per the mass x volume rules defined in Appendix E.

In the case $E / E_B < 1$, $4\pi r^2 b_3$ must be replaced by the actual allowed volume.

The Hydrogen example is shown below.

VI. Mass x Boundary Volume for Hydrogen

We can now more easily visualize the results from Appendix E.

For the Hydrogen ground state n=1 and where the adjacent state is n=2,

E / E_B = .02 (=13.6eV / 680.0eV) kg-boundary (kg-meter^3 using Joule system units of measure.)

As an independent derivation for unit mass (1 kg) x boundary volume for the Hydrogen ground state, we can write (assuming one proton):

$E / E_B = 4/3 \pi (4r_B - r_B)^3 / m_P = .0100$ where

r_B = 5.29E-11 meter,

m_P = 1.67E-27 kg,

and we can ignore m_E.

From Appendix R "Natural Periodic Chart" (attached) and Appendix Y "Neutron Mass" (attached,) we now understand there is one allowed natural isotope of Hydrogen having nuclear mass:

$m_P + m_N = 2m_P$

where m_N does not participate in the strong (K) force with m_E.

Then,

E / E$_B$ = 2 x .0100 = .0200 kg-boundary.

Quantum mechanics is not required to achieve this result.

See Diagrams.

VII. The Nature of Quantum Spatial States (Bound Energy States)

Per Diagrams *The Nature of Spin* and *Forward and Reverse* (attached) from *Cold Fusion*, the atomic nucleus and the electrons of an atom share a dimensional mass-force that determines allowed and disallowed energy states in three dimensions.

Using standard hydrogen (H) as an example, had our history adopted different nomenclature, we could have determined there were three (3) "different kinds" of H, each having a slightly different value of excited energy.

Or, using familiar nomenclature, we could equally determine there were (are) "different types" of electrons, each type having a unique semi-attribute value called "spin." Similarly, we could make up other "particle" attributes, e.g. "color" and so on.

Today, we understand that electron "spin" is an energy state that is a direct function of mass x volume-of-space. (Appendix E, *Changing Your Mind*.)

There is in fact no plethora of different hydrogen (H) atoms and no plethora of "spinning" electrons.

With no understanding of higher dimensional space, and fixated by the perception of "continuous time," our senses have been duped.

Similarly, the so-called atomic "isotopes" shown in Tables 1, 2, and 3 are the same atomic nuclei experiencing multiple energy states. Per Appendix Q, the total atomic mass determines the "space" or radius of the nucleus. Neutron presence represents a variable energy state.

A so-called half-life (of unstable isotopes) is erroneous in its referral to "time," but is accurate as a reference to "change."

For example, per the reference diagrams, an hydrogen (H) electron changes its "fine" or spin-split energy every c / 4 events. It is the same atom and the same electron. Only the energy states change.

VIII. Intersections

Mass is evident only at intersections.

For example, 3-dimensional intersections between 5 and 8-dimensional space (atomic nuclei and subsequent matter, and also black holes,) at 2-dimensional intersections between 3 and 5-dimensional space (the 2-dimensional area of the closed proton mass,) and also the linear (closed) intersection between 2 and 3-dimensional space having the mass dimension 0 (zero.) A point mass (electron.)

These would have the current nomenclature neutron (8,) proton (5,) and electron (3) having respective dimension-of-mass 2, 1, and 0 per the natural sequence.

The various known intersections are represented on the figurative graph "Intersection" extended from the mass-volume graph in Appendix E. The function is in fact hyperbolic.

While the curvature of dimension D through dimension D+1 (Appendix C) is $C = \varphi$, the curvature of an intersection is not that well defined but is bounded.

Boundary condition 1: $C < \varphi$
Boundary condition 2: $C \leq (\varphi + \varphi^2) / 2$ or $C \leq (\varphi + \gamma) / 2$.
Then

$0.5 \leq C_I < \varphi$.

This natural fact is shown in Tables 2 and 3 for the spatial-intersections called "nuclei."

There are no intersections for the curvature $C = \varphi$; instead, that curvature is reserved for the dimension itself and the intersection(s) is gone.

IX. Boundary Energies

In fact (see Table 3,) most elementary nuclei "achieve" energy through the spatial boundary progression unlike hydrogen (H) which loses energy through the spatial progression (the nature of hydrocarbon aging.)

The achieved energy increases as a function of atomic mass or equivalently as a function of increasing E_{CR} as the intersection structures approach the natural curvature limit $C = \varphi$.

From the text (Appendix Q,) nuclear radius increases as a function of nuclear mass (added neutrons) and higher nuclear mass achieves a lower bound nuclear energy.

But the mass is bounded by the mass x volume rules shown throughout the Manuscript and especially in Appendix E.

The additional energy through spatial transition should exactly (statistically) determine the summation of neutron mass in (around) the 3-dimensional intersection (atomic nucleus) between 5 and 8-dimensional space.

X. Closed Spatial Travel

We have proved the smallest r_3 for 3-dimensional spatial closure (C=1) is the Bohr[1] radius r_B per Appendix Z (attached.)

This (smallest) r_3 should in fact represent the largest mass x volume, i.e. for the largest possible black hole.

For the smallest possible mass x volume in 3-dimensions (Hydrogen in the ground state,) r_3 is much larger. Arguably, the largest possible r_3 may represent the 3-dimensional universal "radius."

A radius is a one dimensional distance that represents two dimensions, e.g. the term radius refers to a circle. A circle is a one dimensional "distance" with the curvature C=1 that has closed itself through two dimensions.

While a radius can be integrated to represent a 3-dimensional volume, the radius is in fact a two dimensional reality that is also integrated to represent the added 2-dimensions in 5-dimensional space per Appendix Z.

The possible values for large r_3 are represented by the mass and radius values defined in Appendix G (attached):

$m_H / r_H \leq c^3 / G$ where

G is the known single-dimension gravitational constant, and the subscript H refers to black hole attributes.

We assert this defines the largest possible r_3 radius-of-closure for three dimensional space and represents the radius required for the Hydrogen ground state (n=1) to close upon itself.

The molecular hydro-carbon bound H state should then be closed through a radius (1/4)x (divided by four) the ground state value where the radius is a function of total mass as shown by the expression above.

XI. Closed-Travel Energy

Closed spatial travel, per Appendix Z, for three dimensions through five has the five-dimensional energy:

$E_C = (K m_1 m_2 / r_C^3) \times 2\pi^3 r_C.$

For 5-dimensions pulling on 3,

$m_1 = n_P m_P$ and
$m_2 = n_E m_E.$

For 5-dimensions being pulled by 8,

$m_1 = n_P m_P$ and
$m_2 = n_N m_N.$

Then for the force pulling on *3-dimensional mass*, we can write:

$E_C = (2\pi^3 K) n_P^2 m_P m_E / r_C^2,$

and for the force pulling on *5-dimensional mass*, we can write:

$E_C = (2\pi^3 K) n_P n_N m_P^2 / r_C^2.$

Inversely by r_C^2,

E_C for an atom is small, while E_C for a black hole is large.

Then for a 3-dimensional mass, $E_8 / E_5 = E_{CR}$ where

$E_{CR} = (n_P n_N m_P^2 / n_P^2 m_P m_E)$. Then

$E_{CR} = (n_N / n_E)(m_N / m_E)$ or

$\mathbf{E_{CR} = 1833 \times (n_N / n_E)}$.

XII. n_{MAX}

From Appendix E, n_{MAX} is large but not infinity. Infinity implies the proved falsehood of physical continuity.

E / E_B = Mass x Boundary-Volume

Then,

E / E_B (min) = Mass (min) x Boundary-Volume (min)

where minimum 3-dimensional mass = $1 \times m_E$ and

Boundary-Volume (min) = b_3^3. Then,

E / E_B (min) = $m_E \times b_3^3$.

We can then write:

$n_{MAX} = \kappa c / (m_E \times b_3^3)$ where

m_E = 9.11E-31 kg, and
b_3 = 1.111E-17 meters. Then,

n_{MAX} = **1.3E+91.**

XIII. r_{MAX}, r_C, and E_C

From n_{MAX}, we can now determine r_{MAX} and therefore $r_C = r_C(\text{mass})$.

See diagrams.

$r_{MAX} = c(n_{MAX})^2 b_3 / 2\pi^3$.

From this result, we can now determine E_C.

This is the nature of "time travel" (closed dimensional spatial travel.)

XIV. Inverse and Reverse

We have proved the atomic nucleus is the dimensional "inverse" of a black hole (Appendix Q.)

We assert the "reverse" black hole is a captive three dimensional space "residing" within (surrounded in three dimensions by) five dimensional space.

Five dimensional space has the lowest possible energy per unit mass allowable in three dimensional space, i.e. the largest possible 3-dimensional λ (wavelength or radius.)

The extreme 3-dimensional radius should apply if and only if 3-dimensions become closed (C=1) in five.

To visualize this case, consider the flat plane in the Diagram *3 and 5 Dimensions* "bending or warping" equally in each of the two added dimensions (added from three) that form five dimensional space.

Next, consider the bending to be "complete" so that the two dimensional surface closes itself into a spherical surface.

In this case, 3-dimensions are "closed" (go to no new place) within 5-dimensional space in a similar way to 1-dimension closing itself into a circle through the next higher dimension (D=2.)

XV. Exercise

Exercise 1:

What is the energy E required for a single H_2 gas molecule to return (through one boundary) from 5-dimensional space back to 3-dimensional space? For one mole = 6.0E+23 molecules?

What is the nature of this energy (where does it come from?)

Given:

m_E = 9.11E-31 kg
m_P =1.67E-27 kg

Answer:

One molecule

E_B = 680eV kg^{-1} bnd^{-3}

E = 1.089E-16 (J kg^{-1}) x 2m_P(kg)   (ignoring m_E)

E = 3.637E-43 J

E = 2.273E-24eV

One Mole

E = 2.273E-24eV x 6.0E+23

E = 1.4eV

Nature of Energy

In this case, $E_{CR} = 0$, $C = 0$.

The energy comes from *available* energy state transitions within the molecule itself.

Exercise 2:

Repeat the same exercise for one atom and one mole of ^{56}Fe.

Answer:

One Atom

$E_B = 680 \text{eV kg}^{-1} \text{ bnd}^{-3}$

$E = 1.089\text{E-}16 \text{ (J kg}^{-1}\text{)} \times 56 \text{ (amu)}$ (using 1 amu = 1.66E-27 kg)

$E = 1.012\text{E-}41 \text{ J}$

$E = 6.327\text{E-}23 \text{eV}$

One Mole

$E = 38.0 \text{eV}$

For one atom and one mole of ^{235}U?

Answer:

One Atom

$E = E_{56Fe} \times 235/56$

$E = 2.655\text{E-}22 \text{eV}$

One Mole

$E = 159.3 \text{eV}$

Nature of Energy $= E_{CR}$

Exercise 3:

Consider a perfect and uniform ^{56}Fe sphere having mass M and radius R placed at the surface coordinates $(x_{LGP}, y_{LAP}, z_{R\text{-}Earth})$

1. What is the most convenient frame of coordinates for this exercise?
2. What is the energy requirement E of re-entry to three dimensional space in a single boundary?
3. What are the dimensional curvatures C of the atom? Of the object?
4. What is the range of integers n (as in $\lambda = nb_3$) for ^{56}Fe?
5. What is the closure energy ratio E_{CR} for ^{56}Fe? For the entire object?
6. Could this object close upon itself (run into itself)? In terms of time t, what is the minimum time required for closure? What is the value r_{CS}?
7. Repeat for z having altitude A = 20,000 km.
8. Is there a difference for velocity V (or Omega) in uniform orbit relative to perfectly spherical Earth ground?

Given: Density^{56}Fe = 7,650 kg m^{-3}.

Hint: See the R_C diagram. Also see Appendices E, Q, and W.

Answer: Part 8. No

Exercise 4:

John Doe weighs exactly 100 kg and has M moles of hydrocarbon molecules comprising his living operational molecular groups. Each mole M has the mean value Q hydrogen (H no-neutron) atoms attached in an n=2 molecular bound state. John lives on the surface of the planet Earth.

How many years should John expect to feel healthy given the following:

1. *100% of all H energy transitional states within John are available for transition at every boundary/barrier.*
2. *Statistically, P(%) of John's hydrogen (H) must never fall from an n=2 (molecular bound) to an n=1 state in order for John to feel healthy.*

Given: For H n=2 to n=1, E = hv = 10.2eV.

Answer:

E_B = 680eV kg^{-1} bnd^{-3} .

E_{John} per boundary = 1.089E-16 (J kg^{-1}) x 100 kg = 68,000eV.

With all transitions available,

68,000eV / 10.2eV = 6,667 damaged hydrocarbons per boundary.

Boundaries per "second" = c (=3E+08).

Boundaries per year = $c \times 60 \times 60 \times 24 \times 365$ = 9.46E+15. Then there are

6.31E+19 damaged molecules per year.

Health expectancy = $(1 - P) \times Q \times M / (6.31E+19)$ years.

What if the states were 50% available?_____

XVI. Tables

Table 1—Atomic Examples of the Ratio E_{CR}

Symbol	Nomenclature	Nn/Np	E_{CR} / 1833
H	Standard Hydrogen	0/1	0
^{2}H	Heavy Hydrogen	1/1	1
He	Standard Helium	2/2	1
^{55}Fe	Unstable Light Iron	29/26	1.12
^{56}Fe	Iron	30/26	1.15
^{57}Fe	Iron	31/26	1.19
^{58}Fe	Iron	32/26	1.23
^{59}Fe	Unstable Heavy Iron	33/26	1.27
^{235}U	Uranium	143/92	1.55
^{238}U	Uranium	146/92	1.59

Table 2—Nuclear Examples of Curvature C

Element	Type	AMU	Nn/Np	C Nn/(Nn+Np)
H	standard	1	0	0.000
H	heavy	2	1	0.500
He	standard	4	1	0.500
Fe	unstable	55	1.12	0.527
Fe	stable	56	1.15	0.536
Fe	stable	57	1.19	0.544
Fe	stable	58	1.23	0.552
Fe	unstable	59	1.27	0.559
U	uncommon	235	1.55	0.609
U	common	238	1.59	0.613
Pu	unnatural	244	1.60	0.615
Cm	unnatural	250	1.60	0.616
	unnatural			0.618

Table 3
Nuclear Intersections

Element Nucleus	Type	AMU	Ecr/1833 = Nn/Np	C5-8 Nn/(Nn+Np)	C3-8 (1 - C5-8)	C Int = C3-8 + C5-8	Inter-section	D (Int)	Boundary Energy Req. Earth (eV/kg)	Comment
H	standard	1	0	0.000	1.000	1.000	3-int-5	2	680.0	No 3-D Intersection
H	heavy	2	1	0.500	0.500	1.000	5-int-8	3	-1153.0	
He	standard	4	1	0.500	0.500	1.000	5-int-8	3	-1153.0	
Fe	unstable	55	1.12	0.527	0.473	1.000	5-int-8	3	-1373.0	
Fe	stable	56	1.15	0.536	0.464	1.000	5-int-8	3	-1428.0	
Fe	stable	57	1.19	0.544	0.456	1.000	5-int-8	3	-1501.3	
Fe	stable	58	1.23	0.552	0.448	1.000	5-int-8	3	-1574.6	
Fe	unstable	59	1.27	0.559	0.441	1.000	5-int-8	3	-1647.9	
U	uncommon	235	1.55	0.609	0.391	1.000	5-int-8	3	-2161.2	
U	common	238	1.59	0.613	0.387	1.000	5-int-8	3	-2234.5	
Pu	unnatural	244	1.60	0.615	0.385	1.000	5-int-8	3	-2252.8	
Cm	unnatural	250	1.60	0.616	0.384	1.000	5-int-8	3	-2252.8	
				φ 0.618	γ 0.382	1.000	NA	5	NA	No 3-D Intersection

XVII. Beginning to End of 3-Dimensional Intersections

A Closer Look at Hydrogen
(And Fusion)

From Table 3, we now have a better view of the Hydrogen nucleus.

While the 5-8 three-dimensional intersections (mass-elements) have curvatures bounded by 1/2 and φ, the 3-5 two-dimensional intersection of single-amu Hydrogen has the two-dimensional curvature C=0 with the surface area πr_p^2.

For the H nucleus to truly enter a 3-dimensional (5-8) intersection, it must be closed (C=1) forming a closed spherical surface having surface area $4\pi r_p^2$.

Neutron formation about the H nucleus (causing radius $r_{N2H} = r_p \times 2^{1/6}$ per Appendix Q,) forces a closed nuclear spherical surface.

While the closed nuclear surface has curvature C=1, the three-dimensional 5-8 intersection has curvature C=1/2 as previously shown.

Per Appendix Q, the added mass achieves a lower bound nuclear energy.

The energy difference is easily calculated:

$$\Delta E = \text{GradE}_B \times \Delta r_N = c^3 \times \Delta r_N.$$

This is large and is on the order 10^{10} J.

This type of release is required within a single boundary b_3. The "equivalent time" of release is then on the order 10^{-9} sec.

This is the nature of "temperature."

The release power P_R would be on the order 10^{19} J sec^{-1}.

Per the text *Cold Fusion*, the release wavelengths are large. The type of release above can accurately be called "hot" fusion.

It is then straight forward that the larger nuclei must release energy using smaller wavelengths, and the overall P_R as above must decrease with increasing nuclear mass and size.

Surface Tension

A good example of surface tension is a child blowing a soap bubble from a round (C=1) closed-line plastic toy. The intermediate surface area (C=0) is a thin soap "film."

With a small force of air, the "flat" surface "extends" and then becomes "round" and closes upon itself forming a three-dimensional bubble-volume. It holds itself together until another force breaks its 3-dimensional structure.

The soap bubble is a three dimensional set of events.

In higher dimensional space, forces come from mass and "distance" similar in many ways to the soap bubble.

As per the text **Cold Fusion**, it is constructive to form the surface as with the toy, i.e. over several (many) "frames" of the spatial progression or over the concept of "time."

Forming the surface "instantly" is not recommended.

The soap bubble "requires" energy *to form*. An atomic nucleus, while adding mass, "provides" abundant energy *as it forms*. It is a dimensional soap bubble in reverse.

Instead of low-energy air to form a toy bubble, mass and space are capable of large dimensional energies in a similar way.

As a proof, see the section "The Nature of Mass."

"Once we close our eyes, we can no longer expect to see."

End of the 3-Dimensional Intersections

We now understand the three dimensional intersections more clearly.

From Table 3 and the Diagram *Intersection*, all mass is an intersection having dimension D-minus-2 between the two dimensions of space having dimensions D and D-minus-1.

For example, 3-dimensional intersections are shown to be mass intersections between 5 and 8-dimensional space per the natural (Fibonacci) sequence of growth.

We can now move slightly forward from Appendix Z:

$n_{(p+n)} = (r_H / r_B)^3.$

For a black hole to "disappear" (from 3-dimensional space) then,

$C = \varphi = n_N / n_{(p+n)}$ or

$\varphi \times (m_H / m_P) = n_N$ and

$m_N = \varphi \times m_H$ or

$m_P = m_H (1 - \varphi) = \gamma \times m_H.$

XVIII. Diagrams

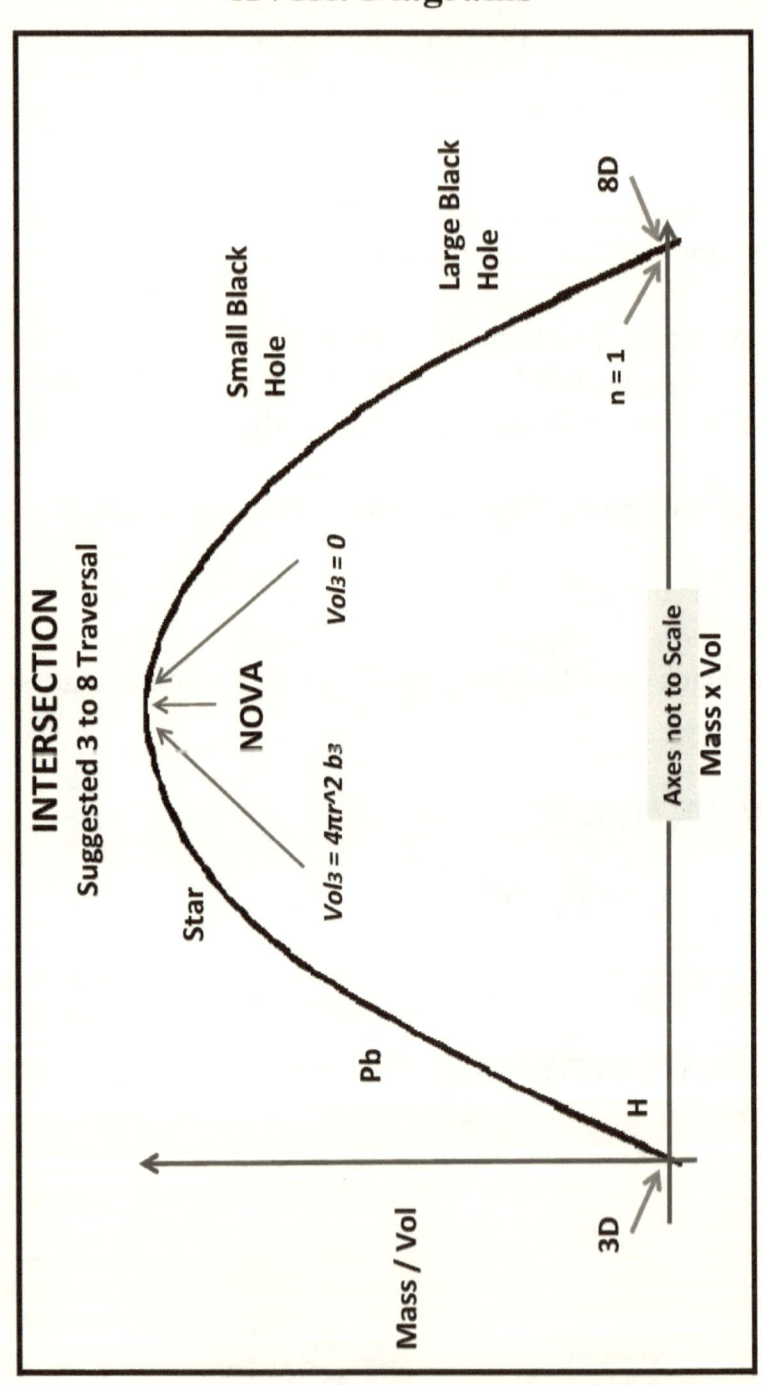

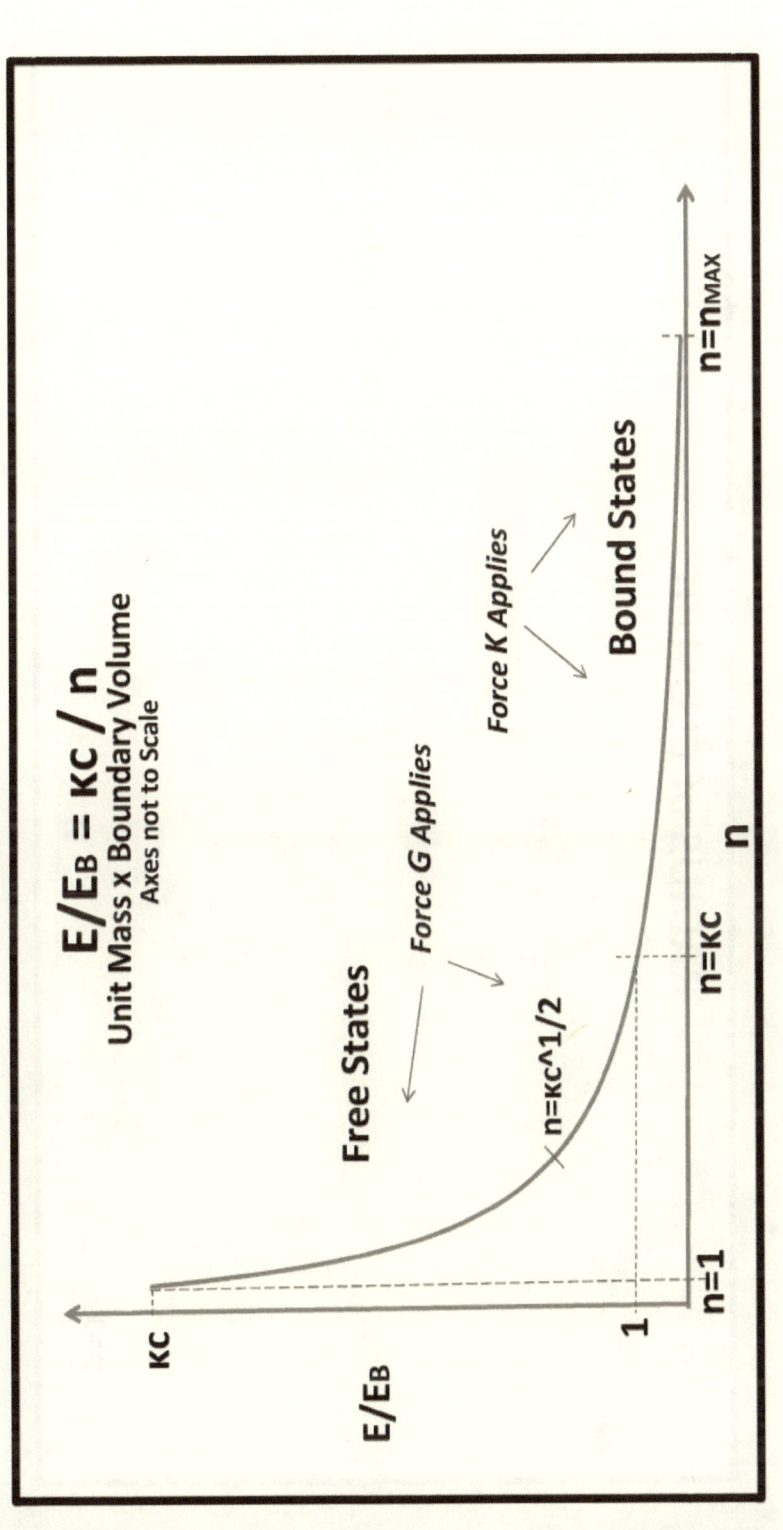

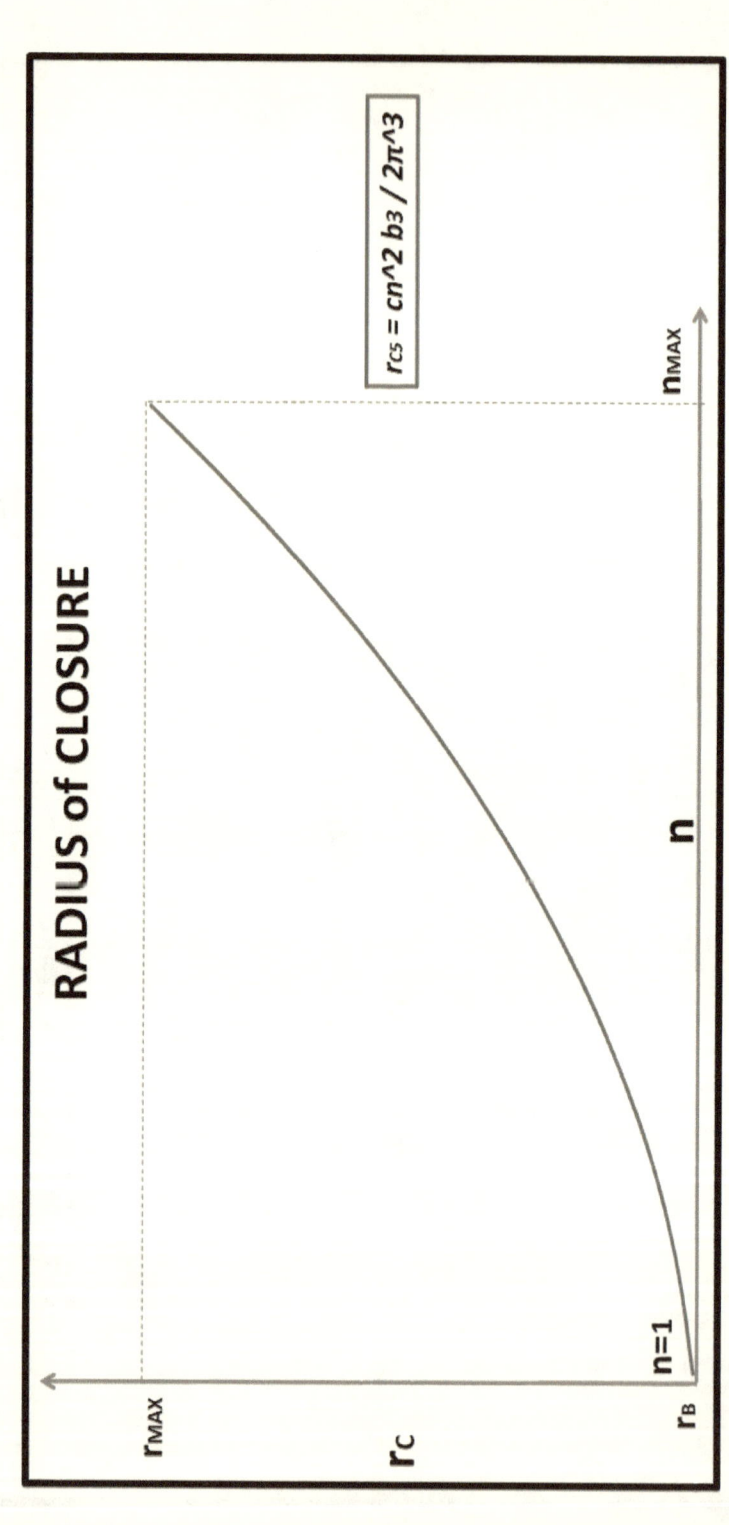

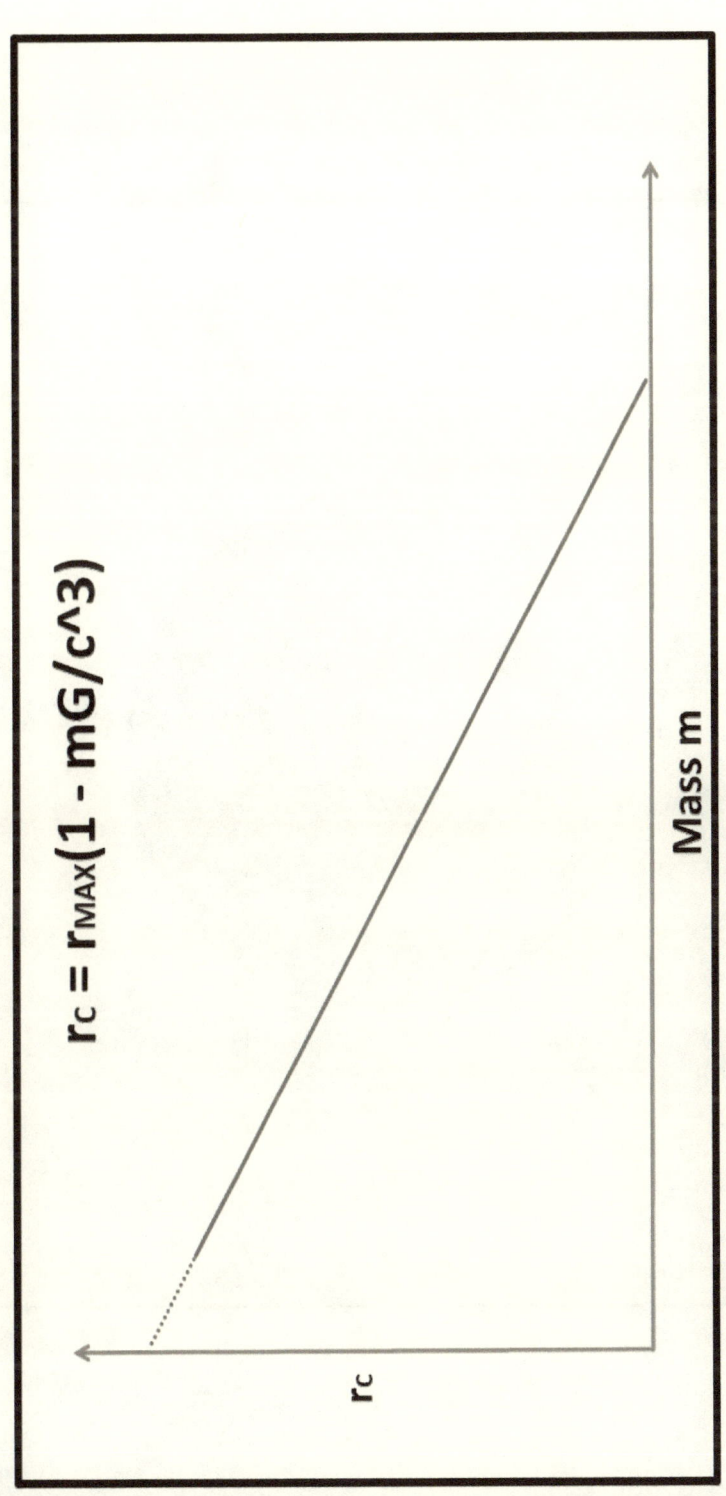

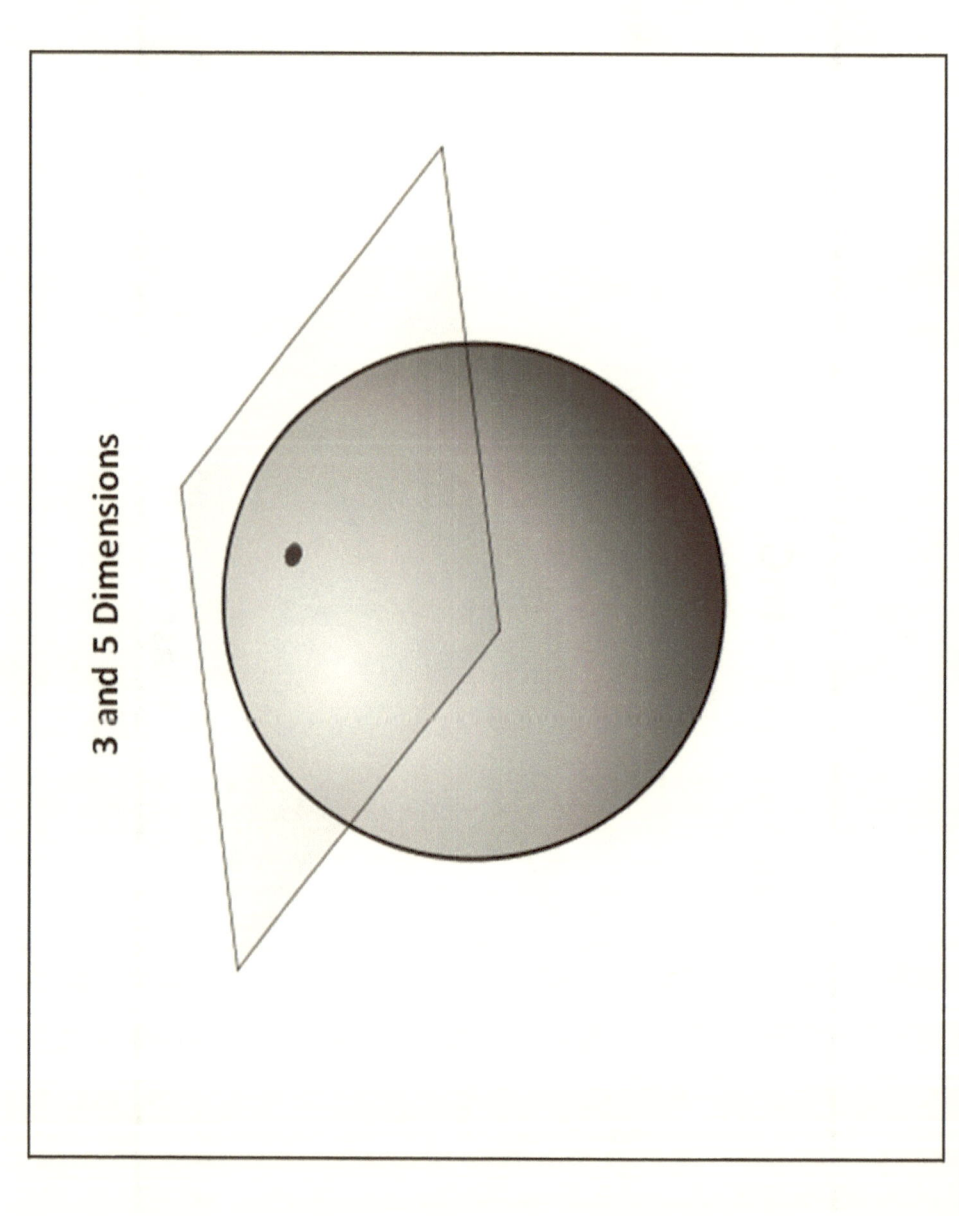

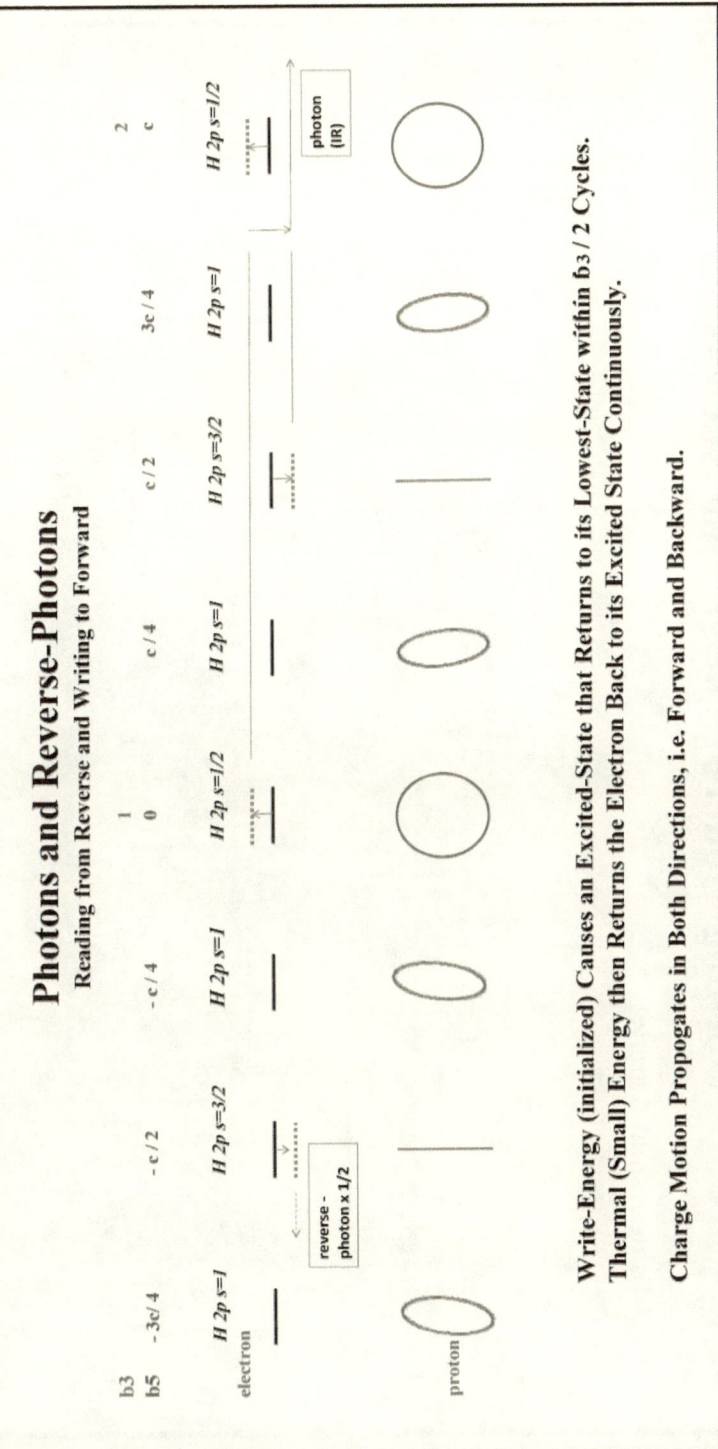

Working Model of The Helium Nucleus

3-Dimensional Atomic Space
$b = b3$

Proton Charges Having
$\delta_{QL} = e+ / 2\pi r$

8-Dimensional Surface Boundary
$b = b8$

5-Dimensional Nucleus
$b = b5$

The Charge Pair has $(1/b5)^3$ Allowed Spin Positions

"Spin" in the Next Higher Dimension

FIVE Dimensions

n = s-1   n = s

SPIN

Nucleus

EVENTS = 1 x c

THREE Dimensions

n = s-1   n = s

Nucleus

EVENTS = 1

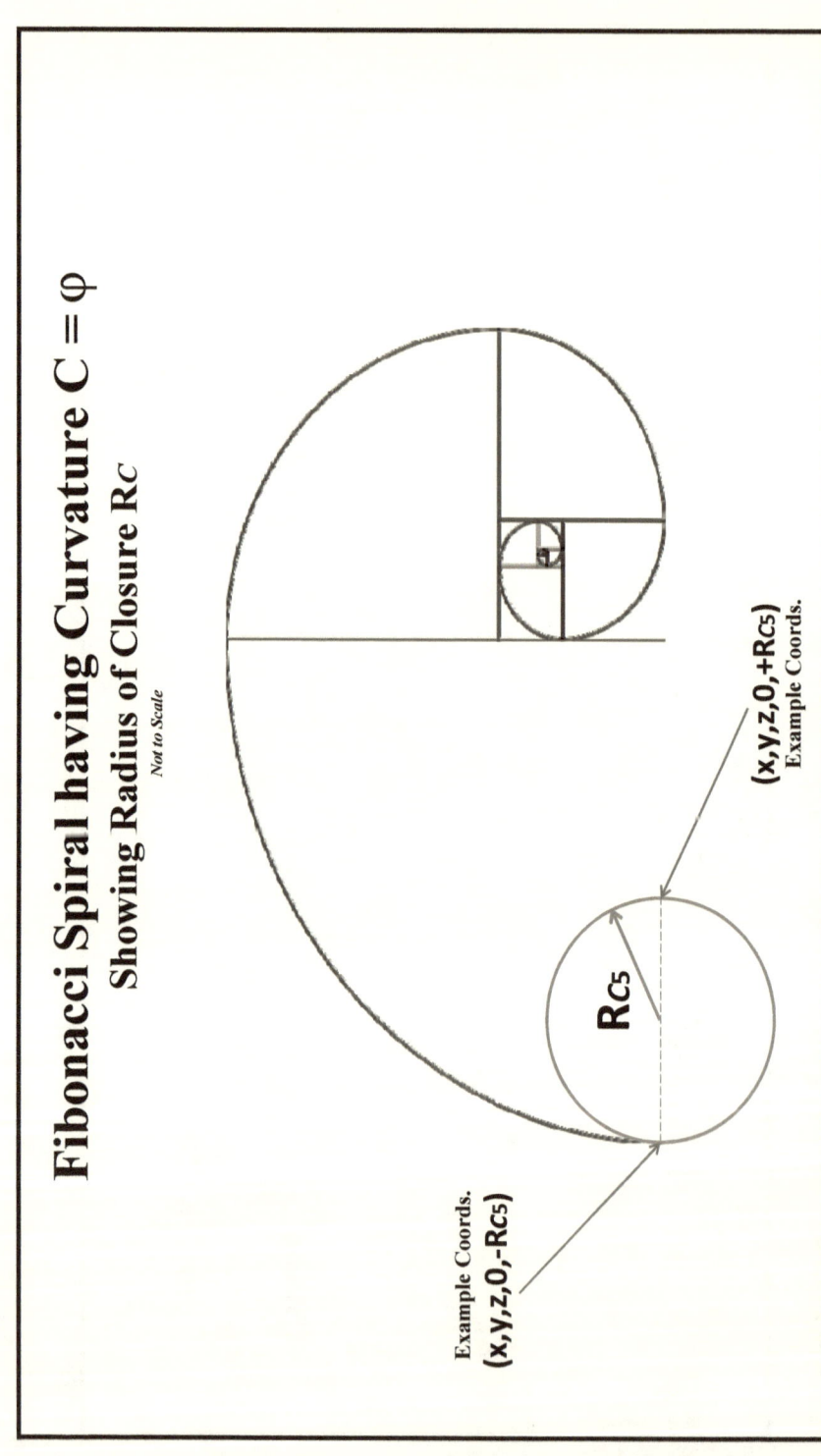

XIX. Technical Conclusions

1. Mass is proved to be a function of space.
2. Atomic nuclei are intersections between 5 and 8 dimensional space similar to black holes.
3. The Periodic Chart of Elements represents mass-space intersections that disappear for curvature $C = \varphi$.
4. Quantum mechanics is a (good) mathematical attempt to force-fit three dimensional thinking into higher dimensional space.
5. Particle physics is an extreme attempt to force-fit the universe into an ill-perceived three dimensional concept.
6. Hydrogen atoms (alone) and hydrocarbon molecules suffer the quantum energetic consequences of $E_{CR} < 1$.

XX. Reference

Manuscript

A Mathematical Transformation of Variables Defining Space-Time and the Constant h

Marc E. King
Silicon Valley, California
June, 2012

Abstract

A variable transformation for time t is supported by wave mechanics and relativity theory and shows that time and space can be related and connected by the concept of physical events per unit space. The transformation confirms our daily macroscopic experience as unchanged from classical physics while still suggests new physics regarding small energies and large spaces. Perceived time can be altered relative to Earth-bound clocks in regions of lower or higher gravitational force. A series of calculation-verifications proves the theory, derives Planck's constant and defines quantum mechanics. Black holes and their mass-radius relationship are defined. The Schwarzschild radius is defined. Minimum and maximum energies are defined.

Introduction

In this model, it is shown that continuous time t and a contiguous view of spatial frames are mathematically the same in the macroscopic sense. A suggested transformation of variables presents interesting differences in concept for small and large energies and spaces.

Concept

We postulate that continuous time as experienced can also be represented, with the same physical result, as a directional spatial sequence or frames of events.

We consider a new unit system using the transformation t = cB with c = speed of light, where one spatial frame (size b) is related to one physical event B by

b (meter) = 1 (event) / B (events meter^-1)

The transformation t = cB implies the units t (sec) = c (met/sec) x B (sec^2 met^-1)

Then B events per meter = B sec^2 per meter, and one physical event = one square second = sec^2.

Derivation

From wave mechanics, we have the Schrödinger equation[i]

dψ/dt = +/- 2πi/h x Eψ as a partial derivative

and the related approximation

ΔxΔk ≥ O(1).

This defines the uncertainty in measurements[ii]

Δx Δp ≥ h/2π.

The Nature of Mass

Implying $\Delta p = m \Delta x / \Delta t$ and using a transformation for Δt, then

$$\Delta p = m \Delta x / \Delta(cB)$$

This leads to

$$m (\Delta x)^2 \geq (h/2\pi \text{ J-s}) (\Delta cB) = (h/2\pi \text{ J}) (cB) (\Delta cB)$$

Per unit mass, then

$$(\Delta x)^2 \geq (h/2\pi)(\Delta cB)(cB) \text{ from transformation.}$$

For a single B (events-meter^{-1}) the corresponding $\Delta x = b$ meters and $\Delta(cB) = 1/(cB)$

Then $b \geq (\text{h-bar})^{1/2}$.

Further defining b-minimum as the minimum Δx and using the positive root in this analysis, then

b(min) =1.027E-17 meters.

Subject to the further justification below, we assert:

$$E = F\text{-sub-B} \times b$$

Where F-sub-B = F-sub-G = the gravitational force at the spatial location of event B.

And on the planet surface, F-sub-B = ma = m x 9.8 meters-sec^{-2}.

Then $E / m = a \times b = 9.8b$ meters $/ (cB)^2$ or

$E/m = 9.8b / (c/b)^2$ and

$E/m = 9.8/9 (10^{-16}) b^3$ J-kg^{-1},

Or we can write the expression:

$E/m/b^3 = 1.089\text{E}{-16}$ J kg^{-1} or

E-sub-B / m = 680 eV / kg for one cubic spatial boundary.

Justification for Spatial Dimension b

Using uncertainty and similar to the derivation above, one estimation using neurological sensory communication as an upper bound on $\Delta x = b$ (in one dimension) for spatial boundary (frame) size is suggested by:

(Spatial Frame Width)$^2 \leq$ (h - bar) x (c) x (Time Required for Sensory Continuity)

Using orders of magnitude 10^{-34} J-sec (and adjusting for units) from wave mechanics and estimating the time required for sensory communication in the range 10^{-3} sec - 10^{-6} sec from synapse switching (potential change) rate, we would then estimate the magnitude:

$\Delta x = b \sim 10^{-14}$ to 10^{-16} meters (for example as an upper spatial bound) in order to perceive continuity from actual contiguity of frames.

The neurological bound approximates the largest frame or spatial size that could be perceived as the continuity of time and accommodates the calculated boundary dimension $\Delta x = b \sim 10^{-17}$ meters.

b(min) = 1.027E-17 meters was derived assuming t = cB so that c is assumed to be the maximum achievable velocity[iii] and as such defines a maximum sequential rate and a minimum allowable b.

Justification for Spatial Barrier Energy

Using a one dimensional example,

E = F x distance.

We are using the transformation t = cB where B = 1 / b and b has the spatial dimension of meters.

The energy associated with the distance b is a function of a force F acting upon a mass m at a particular set of spatial coordinates.

It follows, the innate force acting on the mass m in space is the gravitational force.

There are no external forces to be considered for the mass m for our purpose regarding the transformation associating time and space.

Verification of Calculations

E-sub-B is then a function of gravitational force.

Using the calculations above and for a single event B = 1, we can also write, for the planet surface as an example:

E / m = a x b = b x 9.8 (c / b)^-2 = 9.8 / c^2 / b.

Then $1.089\text{E-}16 = 9.8 / c^2 / b$.

And for the planet surface E-sub-B, we verify our unit of measure calculations:

b meters = (1 event / B events meter^{-1}) = 1.000 as a confirmation of the energy calculation 680 eV / kg.

Independently, we can re-calculate the value of b using F-sub-G on the planet surface:

b = E-sub-B / m x (a)$^{-1}$ = 1.089E-16 / 9.8 = 1.111E-17 meters.

This should be the universal value of b and is independent of F-sub-G since the accelerations "g" cancel for any spatial position.

This value is larger than the allowed minimum calculated b(min) = 1.027E-17 by the difference 8.4E-19 meters and we find the calculated surface value to be approximately 8% larger than the minimum allowed value b(min) using the Earth gravitational acceleration a = g = 9.8 m-s^{-2} and using no transformations in this calculation. We do not pursue further calculations in the present scope. **(See Appendices for calculations.)**

Energy Change as a Function of F-sub-G = F-sub-B

Assuming mass m and boundary b are unchanged, then E-sub-B changes as a function of F-sub-G = F-sub-B the gravitational force at the location of physical event B.

This follows directly from

$E_B = F \times b$.

A smaller gravitational force leads to a smaller E_B relative to the planet surface.

With different E_B, clocks should appear to run at different rates in regions of higher or lower F_G relative to the planet surface.

A fictitious force, like the Coriolis force or the weightlessness of orbit, should not affect the real force $F_G = F_B$.

Conclusions

Continuous time can be represented by a contiguous spatial sequence of frames, or boundaries $\Delta x = b$, while conforming to existing physics in the macroscopic sense and with our sensory perceptions.

The transformation $t = cB$ leads to the spatial frame dimension $b(min) = 1.027E{-}17$ meters and corresponds to 3×10^8 physical events in one second of time t.

For this model,

The surface barrier energy E_B per unit mass = 680 eV / kg has been defined.

E_B is suggested to be a function of gravitational force and so a function of spatial location.

Perceptions of Earth-time and clocks are expected to experience different rates in regions of lower or higher gravitational force relative to the planet surface.

Summary of Results

From Text and Appendices of
Changing Your Mind
A Theory of Space without Time

Per the main text and manuscript, we assert the transformation
t = cB

where t (sec) = c (meter/sec) x B (sec^2 / meter) = c / b (event/ meter) and where a single event = one sec^2 and where we need to remove "time" t from our units of measure for non-macroscopic physics.

It has been shown that

b = 1.111E-17 (meters), and that

E_B = 680eV / 1kg / (b^3 met^3).

The Fibonacci infinite sequence can be described as follows:

Lim (n→infinity) F(n-1) / F(n) = φ = 0.618 . . . and

Lim (n→infinity) F(n-2) / F(n) = γ = 0.382 . . . .

The growth rate of space itself for three dimensions expanding through five dimensions is

r_V = φ^(5-3) so that

V_5 / V_3 = e^(r_V) for one "second" of time t.

Similarly, the growth rate of the natural logarithmic base is

$r_L = (1/\gamma)^{(5/2)}$ so that

$e \rightarrow e^{(r_L)}$ for D=5 traversing through D=8.

Appendix C is attached to this text for reference.

Quantum mechanics is then described as $E / E_B = \kappa c / n$ and

$h = fn(E_B) = bE_B\kappa$.

The definition of quantum mechanics becomes mass x volume and defines the chemistry of the periodic chart of elements, and the Schrödinger² solution is replaced by

$E' = E_B(1 - p/n^2)$.

Dimensional spatial curvatures are defined as $0 \leq c \leq 1$ with the terms

0 = open, and

1 = closed.

Appendix K is attached to this text as a reference for curvatures.

Appendices L and N are attached as references for 5-dimensional geometry and 5-dimensional nuclear forces respectively.

The sequence of spatial progression can be described by the following Appendix P.

Summary of Technical Results for the Text

Fifth Dimension, The Light to See

The 5-dimensional atomic nucleus is defined by the vector:

$\mathbf{Q}_0 = (0, 0, \text{GradE}_B \times r_N^5 / (e+/2\pi)^2, 0, 0)$. Then,

$r_N^6 = b \text{ (mass / charge)} (e+/2\pi)^2 / c^3$, and for a considered 2-amu He nucleus:

$r_{N'} = 3.071\text{E-}14$ meter, and for the stable 4-amu nucleus,

$r_N = 3.447\text{E-}14$ meters.

See Diagrams:

Working Model of the Helium Nucleus, and
Coordinates for the 3-Dimensional Sequence.

Appendix R (also attached) is perhaps the best overall summary of **Fifth Dimension**.

Cold Fusion **Technical Summary**

1. The charge attributes + and - have been shown to be dimensional attributes of mass.
2. The semi-attribute "electron spin" has been shown to represent proton spatial position within the five-dimensional nuclear interior.
3. Quantum mechanics and the dimensional nuclear surface force F_N independently derive a universal gravitational constant.
4. The uniform gravitational constant $K(D) = K(\Delta D) + G$ where $K(\Delta D) = 7.57E+28$ (Nwt kg^{-2} m^2) for $\Delta D = 1$.
5. $\kappa = \gamma' / \varphi'$ defines $h = h(E_B)$ and the transitional value $n = \kappa c$ as well as the transition-time expression $t = (1 + \kappa) / c$.
6. Centers of mass appear to be origins of coordinates for the sequential spatial traversal.
7. The dimensional model $t = cB$ is proved by Appendix W.
8. The five dimensional computer model has further been defined by the material-related value λ.
9. The dimensional cold fusion model has been shown. A figurative example has been suggested.
10. A kernel (CH_3) for the hydrocarbon low energy communication has been suggested.
11. Facts of dimensional neutron mass have been shown in Appendix Q and have been suggested in Appendix Y.
12. The correct nuclear radius values for He and H have been shown.
13. The value ν as in $E = h\nu$ has been redefined and re-unitized in the absence of time t.
14. We now more clearly see the meaning of E / E_B = mass x volume.

15. We now more clearly see the meaning of traversal events B where one second of time t is a representation of c events B.
16. We now understand the perception of time-dilation.
17. We now understand that to move forward we must first turn backward and review Diagram 21.

Appendix B (Reference)

Another calculation for difference b-empirical - b(min) = 7.9%:

In two dimensional physics,

$F = ma = m \times \text{met-sec}^{-2}$ becomes $F_2 = m \times a_2$

$= m \times \text{met} - \text{sec}^{-3/2}$

and one physical event B would no longer have units of $\sec^2$; instead,

$\sec^{3/2}$.

The uncertainty principle then has a transformed h-bar, and now

$b(\min) = (\bar{h})^{1/2} \times c^{1/2} = 1.779\text{E-}13$ meters.

Similarly, $E_B / m = 9.8 / c^{3/2}$ J-kg^{-1} per square boundary = $1.886\text{E-}12$ J-kg^{-1} per boundary and

$b = E_B / m / 9.8 = 1.924\text{E-}13$ meters.

Then we again have the mean factor

$= 1.079$ or 7.9%
between $b(\min)$ and b (from F_G) in two-dimensional space exactly the same as in three-dimensional space.

Appendix C (Reference)

A General Fibonacci Calculation:

The Fibonacci infinite sequence was referenced in Appendix A,

$F(n) = F(n-1) + F(n-2)$ with seed values
$F(0) = 0$ and $F(1) = 1$.

Ratios converge, and

$\lim(n \rightarrow \text{infinity})\ F(n+1) / F(n) = \varphi = (1 + 5^{1/2}) / 2 = .618\ldots$ and

$\lim(n \rightarrow \text{infinity})\ F(n-2) / F(n) = \gamma = .382\ldots$ and so on.

Writing an example expression for spatial dimension ≥ 3 per Appendix A

$$\iiiint\!\!\iiint dV = \iiint\!\!\iiint dV(0) \times \exp(r_V \times t)$$

where $r_V = \text{r-sub-V} = \varphi\verb|^|(D(n+1) - D(n))$

then

$dx / dx(0) = \exp(\varphi \verb|^| (D(n+1) - D(n))\verb|^|1/(n+1))$.

Except we are now doing math in another dimension, and while $e = 2.718$ in three dimensions, the base of natural logarithms should change in higher or lower dimensional space.

For example, in the case of 5 dimensions: $e \rightarrow e^1 / \gamma^1$.

We quickly find dx /dx(0) = 1.08.

A different example, for the case of spatial dimension < 3:

The base e must change as a function of the power of B, i.e. in three dimensional space B ~ sec^2 while in two-dimensional space B ~ sec^3/2.

The difference in power of physical events B

2 - 3/2 = 1/2 and the two-dimensional e = 2.718^1/2.

Then we quickly find dx /dx(0) = 1.08 similar to the previous mean calculations for the difference between b(min) and b-empirical.

The (Fibonacci) calculations hold true for any spatial dimension n moving through n+1 with a dimensional adjustment for e.

Appendix G (Reference)

Density of Matter and Black Holes

As the intersection of one dimensional space (a line) with two dimensional space (a surface) is a single point with zero dimension, the intersection of two and three dimensional space is a line with one dimension, and the intersection of three and five dimensional space should be a surface with two dimensions, then the intersection between five and eight dimensional space should be three-dimensional (observable in 3-dimensions) and is suggested by the spherical volume of a black hole.

From Appendix E, the maximum allowed energy-event is 2.700E+25 J.

Then E-sub-B at a black hole surface should be bounded by the maximum allowed c^3 J kg^{-1}.

For the hole surface:

$E_{Bmax} = G m_H r_H / r_H^2$ or E-sub-Bmax = G x m / r for the hole, and

c^3 = G x m / r relating to the hole, or we can write

$m_H / r_H \leq c3 / G$

or

$m_H \leq r_H c^3 / G$

where $G = 6.673\text{E}{-}11$ met^3 kg^-1 sec^-2,
r_H has units meters, and
c (3E+8 numerical) has units J^1/3.

Then $c^3 / G \leq 2.700\text{E}{+}25 / 6.673\text{E}{-}11$, and

$m_H / r_H \leq 4.046\text{E}{+}35$ kg met^-1

for any black hole.

If we let E-sub-B = $\Delta\lambda$ E-sub-Bmax = $\Delta\lambda\ c^3$ where $0 < \Delta\lambda \leq 1$, and

$\Delta\lambda = nb$ where $1 \leq n \leq 1/b = B$, then

$m_H = r_H\ \Delta\lambda\ c^3 / G$ or

$m_H = K_G\ r_H$ where $K_G = \Delta\lambda\ c^3 / G$.

Appendix R (Reference)

The Periodic Chart of Elements

Then there are two boundary conditions (per Appendix E and Appendix Q) defining the natural elements:

1. E / E_B = mass x volume = atomic mass x $4/3 \pi r_{max}^3$
2. r_N = (nuclear mass / nuclear charge) x (b / Q_0)

where r_{max} is the least bound (closest to E_B) electron atomic energy state radius, and

where r_N is the nuclear radius.

The atom and nucleus must obey both conditions.

While the nucleus achieves a lower bound energy requirement with increasing mass, the atomic mass remains bounded by mass x volume.

Periodic Table of the Elements

Group → ↓ Period	1	2	3	4	5	6	7	8	9	10	11	12	13	14	15	16	17	18
1	1 H																	2 He
2	3 Li	4 Be											5 B	6 C	7 N	8 O	9 F	10 Ne
3	11 Na	12 Mg											13 Al	14 Si	15 P	16 S	17 Cl	18 Ar
4	19 K	20 Ca	21 Sc	22 Ti	23 V	24 Cr	25 Mn	26 Fe	27 Co	28 Ni	29 Cu	30 Zn	31 Ga	32 Ge	33 As	34 Se	35 Br	36 Kr
5	37 Rb	38 Sr	39 Y	40 Zr	41 Nb	42 Mo	43 Tc	44 Ru	45 Rh	46 Pd	47 Ag	48 Cd	49 In	50 Sn	51 Sb	52 Te	53 I	54 Xe
6	55 Cs	56 Ba		72 Hf	73 Ta	74 W	75 Re	76 Os	77 Ir	78 Pt	79 Au	80 Hg	81 Tl	82 Pb	83 Bi	84 Po	85 At	86 Rn
7	87 Fr	88 Ra		104 Rf	105 Db	106 Sg	107 Bh	108 Hs	109 Mt	110 Ds	111 Rg	112 Cn	113 Uut	114 Fl	115 Uup	116 Lv	117 Uus	118 Uuo

Lanthanides

57 La	58 Ce	59 Pr	60 Nd	61 Pm	62 Sm	63 Eu	64 Gd	65 Tb	66 Dy	67 Ho	68 Er	69 Tm	70 Yb	71 Lu

Actinides

89 Ac	90 Th	91 Pa	92 U	93 Np	94 Pu	95 Am	96 Cm	97 Bk	98 Cf	99 Es	100 Fm	101 Md	102 No	103 Lr

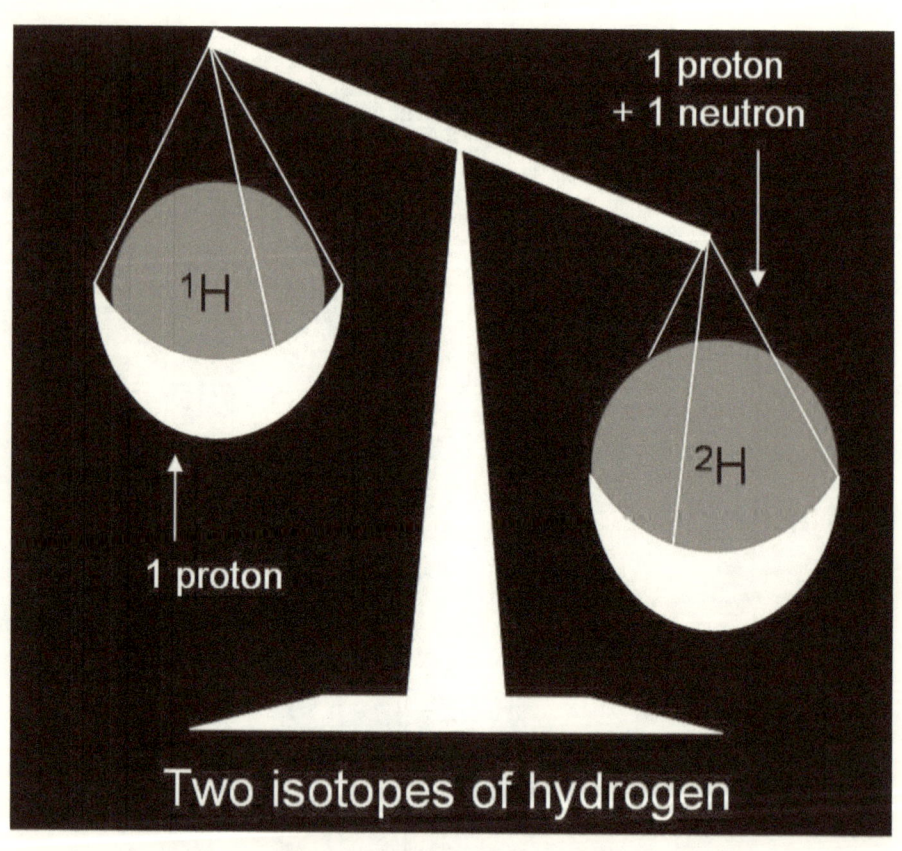

Appendix Y (Reference)

Neutron Mass

As proton mass is five dimensional, we propose neutron mass is eight dimensional. Then,

$m_N = \delta_N \times 4\pi r_N^2 = m_P$.

Per Appendix W, while

m_N, m_P, and m_E all react weakly proportional to G, the pairs

m_E, m_P and m_P, m_N then react strongly proportional to K.

Two neutrons in eight-dimensional space should experience two-dimensional masses weakly acted upon through three-dimensional space,

$F = G m_N^2 / r^5$.

Mass surfaces should then be drawn to "coincide" as a result of this force.

The distance r between the two radii cannot be zero and cannot be smaller than b_{13} per Appendix L.

While G is small, F is large.

A proton and neutron in the same atomic nucleus should interact as shown in Appendix W through the "strong" constant K using the denominator r_N^3.

This strong force should create quantum energy states within the nucleus related to b_s and its own **GradE$_B$**, and so on.

As shown in Appendices Q and R, the neutron mass in the nuclear three-dimensional intersection between 5 and 8 dimensional space must obey the atomic boundary condition mass x volume.

While the nucleus achieves a lower energy state through the addition of neutron mass, the atomic mass x volume remains bounded.

In three dimensions, the masses m_E and m_P are very different.

In five dimensions, the masses m_P and m_N are very different based on the dimensional difference in E_B.

In three dimensions, the difference between m_P and m_N cannot be distinguished.

We suggest that most of the mass that "tips" a three-dimensional measurement scale is five and eight dimensional mass that resides within a 3-dimensional intersection between 5 and 8 dimensional space.

Appendix Z (Reference)

Further Examining Spin and $E = h\nu$

From $t = cB$, we can derive the following per the main manuscript:

$\nu = c / \lambda$ cycles per second.

For the H $2p_{3/2}$ - H $2p_{1/2}$ spin split transition, $\lambda = .027$ meters, then

$\nu = c / \lambda = b\,E+27$ cycles per second, or

$\nu = b\,E+27 / (cB)$ cycles per event. It follows,

$\nu = (b^2 / c)\,E+27$ cycles for the spin split transition, or

$\nu = (b^2 / c)(c^3 / \lambda)$ cycles, and

$\nu = (cb)^2 / \lambda$.

Equivalently, we can write:

$\nu = c^2 / \lambda B^2$. This has the form of a quantum expression.

We can define the maximum possible ν,

$\nu = c^2 / \lambda$.

We can quickly show:

$\nu = c^2 / \lambda = (b_3 c^0 \kappa^6)^{-1} = (b_5 c^1 \kappa^6)^{-1} = (b_8 c^3 \kappa^6)^{-1}$. Then

$n = \lambda \nu \kappa^6$ where $\lambda = nb_3$ per Appendix E.

In this case, $\kappa = \gamma' / \varphi'$ and represents the 3-to-5 dimensional spatial transition per appendices D, E, and W.

Per Appendix Q, a one-dimensional radius in 5-dimensional space obeys a 6th root law.

We do not intend to generalize the spin split calculation and expression for higher dimensional ν (meter^{-1}) in the present scope.

Closed Spatial Travel:

Consider the mass-matter that has the smallest possible natural (not externally excited) |energy| 2-D surface within 3-dimensional space.

This should be the surface having the radius r', where r' represents the 2-D cross-section radius from the single proton nuclear center of an H atom to the ground electron state for the same H atom.

Next, consider the spatial traverse for the single 3-D H (hydrogen) atom through 5-dimensional space:

From the 5-dimensional view, 3-D geometry is closed (curvature=1) in 5-dimensions. Then the 3-dimensional space must traverse a closed 5-D "circle" through the additional two (of the total five) dimensions in 5-D.

Except a 5-D "circle" is not the same as a 2-D circle that can be simply integrated into a closed surface and subsequently the surface into a "volume." While straight-forward, the next volume (5-D) integration involves two dimensions expanding into five.

In 2-D, the relationship for a closed-distance is $d_{C2} = 2\pi^1 r$ because the spatial change is only to-and-from a delta of one dimension. It directly follows, in 5-D the relationship for a closed-distance is $d_{C5} = 2\pi^3 r$.

The minimum allowed distance for the two dimensional surface to progress along the direction (0,0,"v/v") to "return to where it started" or to return "back onto itself" (close itself) is cb_3.

The increase in one-dimensional size (r) from 3 to 5 dimensions is $.08\wedge(5/3)$ per Appendix C.

Then,

$2\pi^3 r(1+.08^{5/3}) = cb_3$ *where*

$c = 2.998E+08$ *events and* $b_3 = 1.111E-17$ *meter-event*$^{-1}$.

Then,

$r = cb_3 / 2\pi^3(1+.08^{5/3}) = 5.29E-11$ *meters* $= r_B$. *In other words,*

$r = r' = r_B$ *the exact Bohr*[A] *radius.*

To generalize:

For multiple mass-pair (m_P, m_E),

$r = cn^2 b_3 / 2\pi^3$,

and as a function of energy,

$r = cn\lambda / 2\pi^3$,

where r is the function $r = r(E/E_B)$.

To continue with the H example:

In 3D, the radius r is not fixed and the 3D geometry is not closed.

In 5D, while $2\pi^3 r_5 = cb_3$ and $s'/s = (1+.08^{5/3})\pi^3$,

in 3D, $s'/s = \pi/\varphi$ or $ds'/ds = \pi/\varphi$.

Arguably, for H:

There is normally no neutron, and the 3-to-5 (requiring a 5-to-8) traverse is not present.

In that case, H can be subject to the small-value G since K is 0 (zero.)

Then H, and H alone, may become a victim of the 3D (G) gravitational energy requirement = E_B = 680eV/kg on the planet surface.

Arguably, were we constructed from 100% heavy-H (as in heavy-water where H is the single one-neutron H isotope,) then the hydrocarbons within us would not "age" because G would not apply in the presence of the stronger force K.

The three-to-five (similar in principle to the 5 to 8) traverse should be an electron transition. These shall be beyond the scope of the present text.

Reference Texts

1. ***Changing Your Mind, A Theory of Space without Time; t=cB.*** Penguin Group, Xlibris Press; 2012. ISBN 978-1-4771-0439-2, LCCN 2012907683.
2. ***Fifth Dimension, The Light to See.*** Penguin Group, Xlibris Press; 2012. ISBN 978-1-4797-0645-7.
3. ***The Modern Health Guide You Cannot Live Without.*** Penguin Group, Xlibris Press; 2012. ISBN 978-1-4797-0643-3.
4. ***Cold Fusion, Dignity of Mind.*** Penguin Group, Xlibris Press; 2012. ISBN 978-1-4797-3139-8, LCCN 20129118957.

www.ingramcontent.com/pod-product-compliance
Lightning Source LLC
Chambersburg PA
CBHW021859170526
45157CB00005B/1881